ELEMENTS OF THE PERIODIC TABLE

WHAT IS HYDROGEN?

KATHLEEN A. KLATTE

Published in 2026 by The Rosen Publishing Group, Inc.
2544 Clinton Street, Buffalo, NY 14224

Portions of this work were originally authored by Linda Saucerman and published as *Hydrogen*. All new material this edition authored by Kathleen A. Klatte.

Designer: Rachel Rising
Editor: Kathleen A. Klatte

Cataloging-in-Publication Data

Names: Klatte, Kathleen A.
Title: What is hydrogen? / Kathleen A. Klatte.
Description: Buffalo, NY : Rosen Publishing, 2026. | Series: Elements of the Periodic Table | Includes glossary and index.
Identifiers: ISBN 9781499478396 (pbk.) | ISBN 9781499478402 (library bound) | ISBN 9781499478419 (ebook)
Subjects: LCSH: Hydrogen--Juvenile literature. | Hydrogen ions--Juvenile literature. | Chemical elements--Juvenile literature.
Classification: LCC QD181.H1 K538 2026 | DDC 546'.21--dc23

Some of the images in this book illustrate individuals who are models. The depictions do not imply actual situations or events.

Manufactured in the United States of America

CPSIA Compliance Information: Batch #CSRYA26. For further information, contact Rosen Publishing at 1-800-237-9932.

CONTENTS

INTRODUCTION

One of the great challenges of human society has always been transportation. How can we move people and goods from place to place, especially over long distances? How can we accomplish this safely, quickly, and economically?

In the 1930s, airships powered by hydrogen seemed like a possible answer. The *Hindenburg* was a dirigible—an airship with a rigid frame that could be steered. It was the largest of its kind at 804 feet (245 m) long. It could carry about 90 people at speeds up to 84 miles (135 km) per hour.

On May 6, 1937, the *Hindenburg* was scheduled to complete a transatlantic passenger run at the naval station in Lakehurst, New Jersey. Tragically, just before landing a fire broke out, igniting the highly flammable hydrogen gas used to inflate the blimp (and possibly other materials). Thirty-six people died, and the enormous ship was destroyed. There were radio and newspaper reporters on hand to cover the landing, so news traveled quickly. Many people considered travel by dirigible to be a failed experiment. The transportation industry turned its attention to airplanes instead.

The *Hindenburg's* outer layer was covered with cotton and other materials. Today's airships use high-tech flame-resistant materials such as Kevlar—the same material that many firefighters' coats are made of.

Today, with dire news about greenhouse gas emissions and climate change filling the headlines, the way we fuel our transportation system is under intense scrutiny. Jet fuel exhaust contains many harmful chemicals that pollute the atmosphere. A possible replacement for fossil fuels is hydrogen. When hydrogen is used for fuel in some ways, the only emission is water, making it a very clean power source.

Until recently, processing hydrogen for fuel and transporting it safely to where it's needed have been challenges. Newer technology makes it possible to process hydrogen from the methane found in natural gas. In this process, carbon dioxide (CO_2) is trapped rather than released into the atmosphere.

Hydrogen is the lightest element—lighter than air. So how do you transport it to where it's needed for use as fuel? People are using new technologies to retrofit existing pipelines with pipe-within-a pipe systems. Hydrogen would move through reinforced pipes set inside a pipe filled with inert gas. The pipes would be equipped with sensors to indicate leaks.

Another option under development is to use hydrogen-powered airships to move cargo—including hydrogen fuel. Airships are once again being considered a viable option for many reasons. New materials such as Kevlar make them considerably safer than the *Hindenburg*. Today's airships are powered with electricity generated by hydrogen-powered fuel cells. This hydrogen fuel produces zero carbon emissions, and the hydrogen used for lift is less expensive than helium. Airships can lift off and land without the use of runways, which take up a lot of land and need special maintenance. Although airships don't move as fast as planes, they're faster than cargo ships.

Nearly a century after a hydrogen-filled ship caused one of the most famous transit disasters in history, hydrogen is once again in the spotlight—this time as a potential source of clean, renewable fuel for transportation and other uses. Read on to see how this lightest of elements pulls its weight in the natural and industrial worlds.

At this time, most retail hydrogen fuel stations are located in California.

CHAPTER 1

H IS FOR HYDROGEN

The periodic table of elements begins with hydrogen (H). It's the simplest and most abundant element in the universe. An element is a substance composed of only one kind of atom. The nucleus of a hydrogen atom contains one proton and one electron. In its gaseous state, two hydrogen atoms come together to form a diatomic molecule (H_2). Hydrogen gas can't be seen, smelled, or tasted. Pure hydrogen forms less than 1 percent of Earth's crust. It's most commonly found as part of the compound that forms water molecules (H_2O)—and water covers three-quarters of Earth's surface. There is hydrogen in almost everything around you. If you look up at the night sky, there's hydrogen in all those stars!

A British scientist named Henry Cavendish (1731–1810) is credited with discovering hydrogen. In 1766, he published a paper about gases produced during experiments in a laboratory. One of these was a substance he called "inflammable air." This was produced by dissolving zinc in acid and capturing the resulting gas in a bottle.

The word "hydrogen" was coined by a French chemist named Antoine Lavoisier (1743–1794). "Hydrogen" comes from the Greek words *hydro*, meaning "water," and *genes* (pronounced "GEN-us"), meaning "creator." Lavoisier was influential in defining the modern science of chemistry. He kept detailed records of his experiments. His wife drew pictures of the tools he used so other scientists could replicate his experiments. He helped create the nomenclature, or system of naming chemical elements, that's still used today. He also published a paper explaining how chemists should conduct experiments and explain their results.

Antoine Lavoisier was part of a new generation of scientists. Rather than just proposing theories and discussing them, he tried to prove or disprove them through rigorous, carefully recorded experiments. His particular interest was called pneumatic chemistry—the study of gases.

ATOMS AND ELEMENTS

After hydrogen was discovered, scientists realized that hydrogen is the simplest of all atoms and that it also is the most prevalent, or widespread.

All elements are made up of atoms. In fact, each element is made up of only one kind of atom. So, whether you have one atom of gold or a nugget made up of many atoms of pure gold, you still have just one element—gold (Au).

No two elements are exactly alike. This is because of their atoms. Atoms unite with other atoms to form molecules. A molecule is two or more atoms that are chemically bound together in a fixed ratio, or number. A fixed combination of atoms will create different hydrogen molecules. One hydrogen atom is called atomic hydrogen (H). Hydrogen gas (H_2), or molecular hydrogen, is made of two hydrogen atoms bound together to make a diatomic molecule.

In order to see atoms, you need to use an electron microscope. These powerful devices can magnify an image up to 500,000 times its actual size!

THE MAN WHO WEIGHED THE WORLD

Henry Cavendish was a great-grandson of the second Duke of Devonshire. He studied at Cambridge University, although he didn't complete a degree. He lived in London, England, with his father in a house with stables that had been transformed into a laboratory. He attended meetings of the Royal Society with his father and was made a Fellow of the Royal Society in 1760.

Cavendish published 16 papers in the journal of the Royal Society. He was awarded the Copley Medal for his papers on "factitious air," which included his findings about the substance we now know as hydrogen. He's also famous for a series of experiments to determine the mass of Earth. His results are considered very close to the real number by experts working with today's technology. Cavendish probably made more discoveries than he published, as he was a very shy person.

The most important particles that make up atoms are subatomic particles known as protons, neutrons, and electrons. Protons have a positive electric charge, neutrons have no charge, and electrons have a negative electric charge. All atoms contain a nucleus, or center, where the protons and neutrons are contained. Outside the nucleus are shells that contain the electrons. The first shell can hold only two electrons, the second shell can hold up to eight, and the third shell can hold up to 18.

A hydrogen atom (H) is very simple—it consists of one proton and one electron. A hydrogen molecule (H_2) consists of two bonded hydrogen atoms.

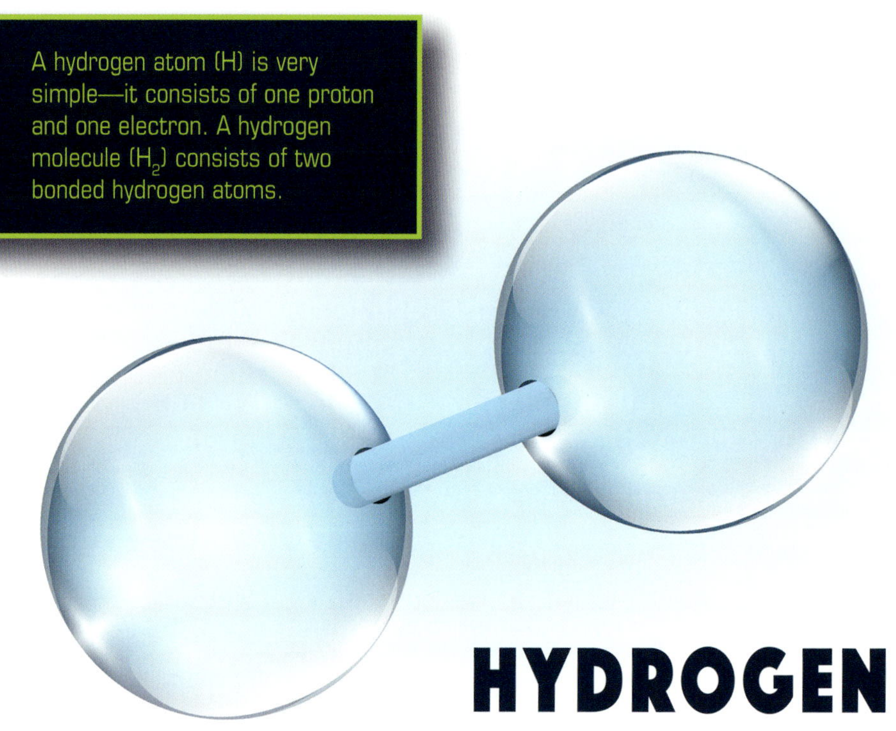

HYDROGEN

A hydrogen atom contains one electron, one proton, and no neutrons. The number of protons usually determines the element's atomic number. Each element's atomic number is unique. The periodic table is organized by atomic number from lowest to highest. Therefore, hydrogen, with an atomic number of one, is the first entry.

Hydrogen also has the lowest atomic weight, or relative atomic mass, of any element. The atomic weight is the mass of one atom of the element. Atomic mass is measured using a technique called mass spectrometry. A hydrogen atom weighs 1.00784 atomic mass units. An atomic mass unit (amu) is defined as 1/12 the mass of a single carbon-12 atom.

Generally, an atom has the same number of electrons and protons. Some atoms of the same element have different numbers of neutrons. These are called isotopes.

The periodic table is constantly being revised as new research defines more elements. The organization that approves the inclusion of new elements and standardizes naming is the International Union of Pure and Applied Chemistry (IUPAC). The 2022 edition of the periodic table recognizes 118 elements. More than 90 are naturally occurring, and the rest are man-made.

The sun is made up of about 3/4 hydrogen, 1/4 helium, and traces of other elements. The process of nuclear fusion converts hydrogen into helium. This means that over time the sun will contain less hydrogen and more helium.

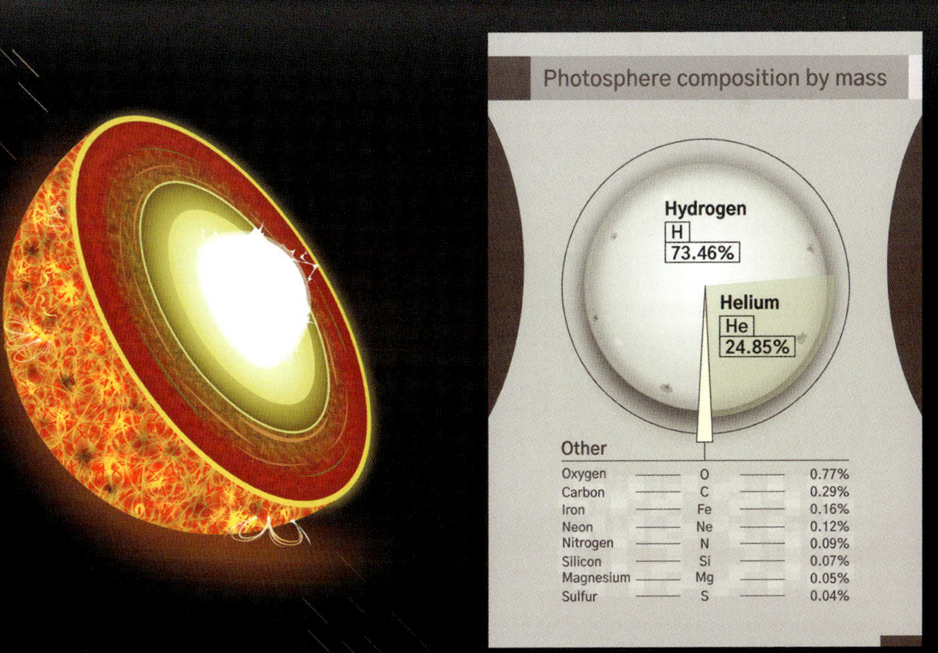

CHAPTER 2

MENDELEEV'S TABLE

The periodic table of elements didn't always look the way it does now. It was first envisioned by a man named Dmitri Mendeleev (1834–1907). The first version of the table was presented in 1869. It contained 63 elements. Mendeleev believed that arranging the known elements in order of their atomic weight showed clear similarities in groups of elements. Scientists of the time agreed that there were probably more elements than those currently identified. Mendeleev theorized that those elements would fit into gaps left in his table. He was proven correct during his lifetime, as more elements were identified and found to fit the gaps in the periodic table.

The modern periodic table is arranged by atomic number instead of atomic weight. The atomic number of an element is the number of protons found in an atom. Hydrogen's atomic weight and atomic number are the same—one, making it the lightest element—so it remains in the first spot on the table today.

HYDROGEN SNAPSHOT

CHEMICAL SYMBOL: H

PROPERTIES: Colorless, odorless, tasteless, and extremely flammable

DISCOVERED BY: Henry Cavendish, England, 1766

ATOMIC NUMBER: 1

ATOMIC WEIGHT: 1.00784 atomic mass units

PROTONS: 1 **ELECTRONS:** 1 **NEUTRONS:** 0

DENSITY AT 273 K: 0.0899 g/L

MELTING POINT: -434°F (-258.975°C), 14.025 K

BOILING POINT: -422.99°F (-252.77°C), 20.38 K

COMMONLY FOUND: Water, air, Earth's crust, sun, and stars

It is important for scientists around the world to use the same system of naming elements. It makes it easier to share their findings.

Elements are then organized into periods and groups. Periods are horizontal rows in the table. In each period, elements are arranged by increasing atomic number, reading from left to right. Since hydrogen's atomic number is 1, it sits in the upper-left corner of the table.

Groups are assigned by the number of electrons found in the outermost shell of the atom. The number of the group appears above each column of the table, numbers 1 through 18. These groups are sometimes designated with Roman numerals and the letter A or B. The elements within a group have similar chemical properties and are sometimes referred to as a "family" of elements. The elements in the A groups are known as representative elements. Elements in the B groups are transition elements

Many of the groups of elements have a group name. For example, elements in group IA, where hydrogen is located, are called alkali metals. However, despite its position on the periodic table, hydrogen is not considered a metal like the rest of the group. Even though hydrogen is in the same group as the alkali metals, the first alkali metal is lithium (Li), followed by sodium (Na). They continue all the way down to the heaviest alkali metal, francium (Fr). Alkali metals are metals that can combine easily with other elements. Alkalis are also very malleable (they can be extended or shaped by hammering), ductile (they can be pulled into wire), and good conductors of electricity. Alkali metals react violently when they are mixed with water.

Hydrogen is a nonmetal, which is pretty much the opposite of an alkali metal. Nonmetals are not good conductors of electricity, are very brittle, and are not ductile. Nonmetals exist as gases, such as hydrogen, and solids, such as carbon.

Shown here is Mendeleev and his periodic table. Today, most scientists think all naturally occurring elements have been identified. Because the current periodic table is organized by atomic number, it has no gaps. Man-made elements have higher atomic numbers and fit into spaces at the end of the table.

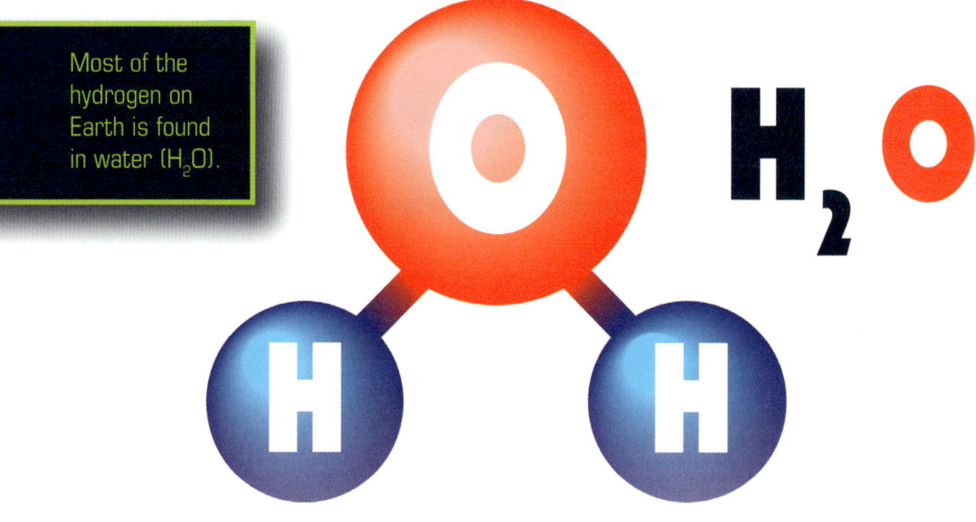

Most of the hydrogen on Earth is found in water (H_2O).

Elements in certain groups have similar chemical behaviors because of their electric forces. If you've ever held two magnets together and felt the strong force that makes it impossible for you to stick the two magnets together, then you have an idea of what happens when two protons are near each other. Because protons are positively charged, they repel each other. But it is the negatively charged electrons surrounding the nucleus that really determine how elements behave and react with one another.

Knowing that hydrogen's atomic number is 1, we also know that a single hydrogen atom has just one proton and one electron. Having just one electron makes it easy for this element to combine with other elements to form compounds. Compounds are made of elements that have bonded together. For example, when hydrogen combines with oxygen, it becomes H_2O, which is water. This is why hydrogen was dubbed the "water generator."

Another interesting characteristic of hydrogen is that it commonly appears in the form of a diatomic molecule, which is two atoms of hydrogen bonded together. When looking at a chemical formula, hydrogen is represented by H_2. The "2" tells us that there are two hydrogen atoms within every diatomic molecule of hydrogen. Looking at the chemical formula for a water molecule (H_2O), we can see that there are two atoms of hydrogen (one diatomic hydrogen molecule) and one atom of oxygen.

MAKING HIS MARK ON HISTORY

If you want to have a school named after you, you either donate a lot of money to education or you do something amazing. Dmitri Mendeleev was not from a wealthy family, so his contribution to learning came in the form of his chart that would change the field of chemistry—and ultimately lead to a university in Russia being named in his honor.

Mendeleev's contributions to chemistry are also remembered with a man-made element named in his honor—mendelevium (Md). Its atomic number is 101, and it was discovered in 1955.

The elements discovered today are often the result of experiments conducted in high-tech laboratories using radiation.

Mendeleev was born in Siberia, a very cold region in Russia, and was the youngest of 14 living children. After studying and teaching in St. Petersburg, he traveled to Paris, France, to get his doctorate. Along the way, stories say, he stopped at some salt mines in Poland, and it was there that he first had the notion of what would later become the periodic table. Mendeleev looked around the salt mines and wondered how sodium, chlorine (Cl), bromine (Br), potassium (K), and other elements present in the mine behaved or reacted with each other. This idea stuck with him and presented itself several years later while he was teaching in St. Petersburg. He wanted to develop a way for his students to see properties of elements at a glance and was working on arranging and rearranging the known elements. Some accounts say that the arrangement of the elements by atomic weight came to Mendeleev in a dream. Whether or not he was sleeping when he thought of it, Mendeleev's periodic table certainly was an eye-opener for the scientific community when it was presented in 1868 and published the following year.

Mendeleev's arrangement of the elements has been altered and rearranged as scientists have made new discoveries. But its foundation of providing clarity to a confusing world of elements has withstood the test of time.

HEAVYWEIGHT CHAMPIONS

New elements are still being discovered and added to the table today. Generally, the new elements are man-made. To be added to the periodic table, new elements have to undergo rigorous testing. Different scientists in different labs need to be able to replicate the same results. The latest additions to the periodic table are four superheavy elements created in laboratories. In 2016, elements 113, 115, 117, and 118 were added to the table.

ATOMIC NUMBER	INITIAL DESIGNATION	OFFICIAL NAME
113	UNUNTRIUM (UUT)	NIHONIUM (NH)
115	UNUNPENTIUM (UUP)	MOSCOVIUM (MC)
117	UNUNSEPTIUM (UUS)	TENNESSINE (TS)
118	UNUNOCTIUM (UUO)	OGANESSON (OG)

CHAPTER 3

HOW OLD IS HYDROGEN?

Most atoms in the universe—about 90 percent—are hydrogen. Hydrogen is all around us—in the water that covers much of our planet and the sun that shines in the sky. It makes up about two-thirds of our bodies. But where did it all come from?

Today, most scientists believe that the universe began about 13.8 billion years ago in an explosion commonly referred to as the Big Bang. Hydrogen and helium are the lightest elements in the universe, and they're also believed to be the oldest. Scientists believe that the first hydrogen and helium atoms formed about 380,000 years after the Big Bang. That's how long it took for the plasma from the initial explosion to cool enough to begin forming atoms.

About 400 million years after the first hydrogen atoms appeared, clouds of hydrogen gas began to form stars like our sun. And about 4.4 billion years ago, the surface of Earth cooled enough for water to begin to accumulate on its surface, paving the way for life as we know it.

The Big Bang started billions of years ago. But when did it end? Scientists think it's still going on. Using high-powered telescopes, scientists can observe that the universe is still growing!

Because hydrogen is so light, Earth doesn't have a strong enough gravitational pull to hold it in the atmosphere. However, there are other planets in our solar system with hydrogen-rich atmospheres. These include Jupiter, Saturn, Uranus, and Neptune.

A GLASS OF HYDROGEN

If there's very little hydrogen in the air, then where is this supposedly abundant element? Usually, hydrogen is found with oxygen in the form of water in lakes, streams, seas, and our vast oceans. So, every time you get a glass of water, take a shower or a bath, or dive into a pool, you are experiencing hydrogen.

If we boil water, will that separate hydrogen from oxygen? No. Boiling water just creates steam, which is the gaseous form of water. However, electrolysis can turn water into hydrogen. Electrolysis is when an electric current runs through a nonmetallic electric conductor such as water. An electric current through water will break the bonds between hydrogen and oxygen, separating the two elements.

Other ways to release hydrogen are through the steam from heated carbon; the decomposition of certain hydrocarbons; the reaction of sodium or potassium hydroxide on aluminum; and the displacement (removal) of hydrogen from acids by certain metals. In the United States, people use these methods to produce hydrogen. The hydrogen is then used as a source of energy in nuclear power plants and to fuel spacecraft.

Hydrogen is also found in Earth's crust. But there isn't a lot there—only about 3 percent of the crust is hydrogen. The rest of the crust is made up of oxygen (47 percent), silicon (Si; 28 percent), aluminum (Al; 8 percent), and a variety of other elements, including iron (Fe), calcium (Ca), and sodium (Na).

Car makers are developing vehicles that can be powered with clean hydrogen fuel. As of 2023, there were 59 retail hydrogen fuel stations in the United States.

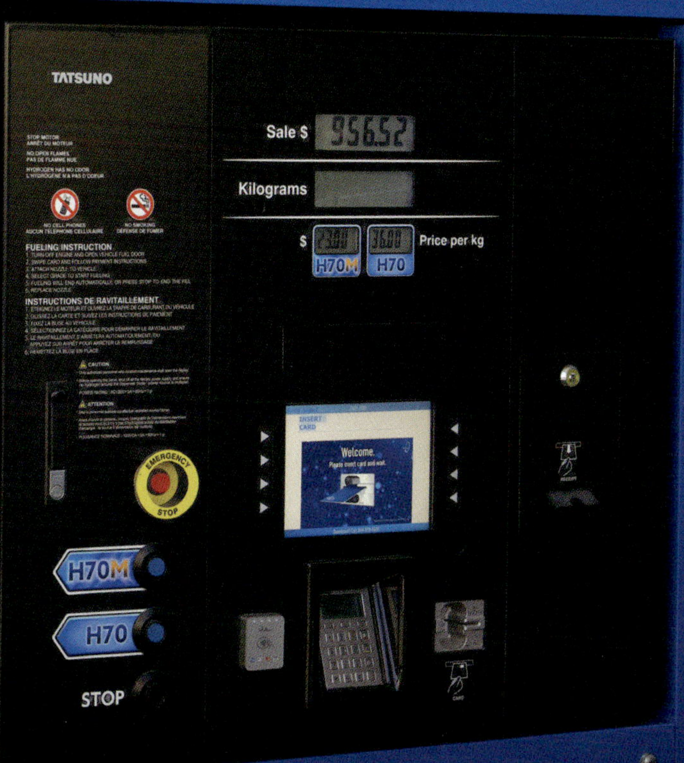

Hydrochloric acid is one of the most corrosive substances in the world. Yet it's produced in your stomach, and without it, you couldn't digest the food you eat.

LET'S MIX IT UP!

Hydrogen, especially in water, plays an important role in the relationship between groups of chemical compounds called acids and bases. When acids and bases react with each other, they produce completely new substances.

When dissolved in water, acids will produce hydrogen ions. Natural acids are found in fruits such as lemons and have a sour taste. For instance, lemons contain a high concentration of citric acid, the chemical that makes your lips pucker. Citric acid mostly is a safe acid, but some acids can burn skin or tissue. Hydrochloric acid (HCl) is a dangerous acid that can be strong enough to corrode metal. However, in a weakened form, hydrochloric acid can be helpful. In fact, a diluted form of hydrochloric acid is found in your stomach and aids in digestion. Other acids are nitric acid (HNO_3), sulfuric acid (H_2SO_4), and perchloric acid ($HClO_4$).

Bases, when dissolved in water, will produce hydroxide ions (OH-). Some examples of bases are ammonia (NH_3), lithium hydroxide (LiOH), and sodium hydroxide (NaO). Bases taste bitter, but you would not want to taste-test a base. Bases denature (damage) proteins, and proteins are what you are made of.

HOW MUCH HYDROGEN?

You, your teacher, your dog, the trees in the park, the birds in the trees, the worms in the soil—every living creature contains hydrogen. In fact, there are about 1.6 ounces (45.4 g) of hydrogen for every 1 pound (454 g) in a human being. So, if you weigh 100 pounds (45.4 kg), you contain 10 pounds (4.5 kg) of hydrogen. With hydrogen making up about 10 percent of our bodies, it makes sense that it is called one of the building blocks of life. Other building blocks include carbon (C), oxygen, nitrogen, sulfur (S), and phosphorus (P). All these elements combine easily with hydrogen.

Tiny hydrogen atoms are the building blocks of everything that's ever lived on Earth.

ALL ABOUT ISOTOPES

An isotope of an element is an atom that has the same number of protons but a different number of neutrons. Hydrogen has three different naturally occurring isotopes. It's the only element with names for its isotopes—protium (H), deuterium (D), and tritium (T).

- Protium is the common version of hydrogen and is represented by the letter H. Protium has one proton and no neutrons. It's a stable isotope, meaning it's not radioactive. The name "protium" isn't commonly used.
- Deuterium has one neutron and one proton. The name comes from the Greek word *deuteros*, meaning second. It was discovered in 1931 by an American chemist named Harold C. Urey (1893–1981). He won a 1934 Nobel Prize for his discovery. Deuterium is also a stable isotope—it's not radioactive and is not considered toxic. When deuterium bonds with oxygen, it forms "heavy water" (D_2O).
- Tritium has one proton and two neutrons. It's an unstable or radioactive isotope. Tritium is formed in the upper atmosphere by exposure to cosmic rays. It can fall to Earth in rain. The tritium in radioactive rain isn't considered harmful because there's so little of it and because the radioactivity decays quickly.

Naturally occurring hydrogen is a mix of all three isotopes. It's about 99 percent protium, less than 1 percent deuterium, and trace amounts of tritium.

In addition to the three naturally occurring isotopes, there are four that are man-made. They exist only in labs and are created through ionizing radiation, one of the two kinds of radiation explained below:

- Ionizing radiation emits electrons from an atom. This creates a very unstable substance, sometimes so unstable that it explodes. The electrons that leave the atom are very dangerous. They can damage living cells and are known to cause cancer.
- Non-ionizing radiation, such as what is experienced from heat, visible light, or radio waves, is not strong enough to cause electrons to be forced out. This radiation is not considered dangerous.

Harold C. Urey won a Nobel Prize for his discovery of deuterium. He also worked on the Manhattan Project during World War II and NASA's Apollo program.

CHAPTER 4

HYDROGEN AND FRIENDS

Most of the hydrogen on Earth exists as part of compounds. That is, it's bonded to other elements to form a different substance. The most familiar hydrogen compound is H_2O—water, which is a simple compound formed by two hydrogen atoms and one oxygen atom.

However, some hydrogen compounds are very complex. Hydroxyzine ($C_{21}H_{27}ClN_2O_2$) is a compound that is made up of 21 atoms of carbon, 27 atoms of hydrogen, one atom of chlorine, two atoms of nitrogen, and two atoms of oxygen. Hydroxyzine is a drug that combats the effects of histamine—a natural substance that causes allergic reactions such as hives. Hydroxyzine can also make people relaxed and sleepy, so it's also used to treat anxiety and sleeplessness.

Hydrogen is found in many, many compounds. Scientists estimate that there are more carbon compounds than any other kind, but most of them contain hydrogen. Carbon compounds containing hydrogen are found in all plants, animals, and even in fossil fuels like gas and oil.

A BOTTLE OF BUBBLY

Hydrogen combines with oxygen to make water, but it also teams up with oxygen to form another, bubblier compound. If you have ever fallen on the sidewalk and scraped your knee, then you might have experienced the compound of two hydrogen atoms and two oxygen atoms—H_2O_2. This compound is hydrogen peroxide. In its most watered-down state, hydrogen peroxide is a bubbly disinfectant that fights germs and is often found in first-aid kits. It helps cuts and scrapes heal better.

Hydrogen peroxide has long been a staple of first-aid kits. However, recent research has shown that soap and water is just as effective at washing away germs and is gentler for your skin.

Hydrogen peroxide will also bleach items, so you don't want to spill it on your clothes! Some hair-lightening kits contain hydrogen peroxide, and sometimes the chemical will cause hair to turn more of a brassy orange than the blond color that was intended. In highly concentrated solutions, hydrogen peroxide can be found in rocket fuel. H_2O_2 is pretty amazing—not only can it turn your hair orange, but it can also shoot you into space!

WHAT'S THAT SMELL?

While hydrogen by itself does not have an odor, when it combines with sulfur, you can definitely smell it—but you wouldn't want to! Hydrogen sulfide (H_2S) is flammable, smells like rotten eggs, and is found in mineral waters, volcanic gases, and decaying matter. In fact, the decaying matter in your stomach can create hydrogen sulfide gas, which can be expelled as flatulence.

The natural gas used for heating homes and businesses is naturally odorless. Utility companies add the smell of sulfur so people will notice when there's a gas leak.

Cyanide is a deadly poison. However, it's used in minute amounts in drugs for high blood pressure.

DEADLY CYANIDE

Another hydrogen compound that has an odor is hydrogen cyanide (HCN). This poisonous and deadly compound smells like almonds. Hydrogen cyanide can be found in the production of fuel from coal. The compound can also be used to make plastic. If you've ever eaten a whole peach and have been left with just the pit, also called a stone, then you've been in contact with something that can become naturally occurring HCN. But don't worry, it is perfectly safe in peaches—just don't try to eat the pit.

HYDROGENATION: GOOD OR BAD?

Besides providing us with something to drink or making sure a cut doesn't get infected, hydrogen is also used to process food. Processed food is treated or prepared using artificial means.

Whenever you spread margarine on your toast, you're getting ready to eat a result of the process called hydrogenation. The hydrogenation process works like this: When hydrogen is added to the double bonds of carbon found in animal fat or vegetable oil molecules, it hardens those molecules. This hardening makes it possible for a stick of margarine to sit on the dining table and not turn into a plate of yellow ooze. But if you've ever seen the gooey mess that happens if you accidentally leave margarine out for too long in the heat, then you know that the hardening of the molecules can be reversed by a rise in temperature.

Hydrogenation hardens oils, resulting in a product called trans fat. This helps prevent food products from going bad. This sounds like a good idea, doesn't it? Trans fat used to be very common in many foods people like to eat, including french fries, cookies, and potato chips. Now, trans fat has been linked to heart disease and other illnesses and artificial trans fats have been banned in the United States. Many companies now manufacture the foods people love without using trans fats. Frito-Lay, the maker of Doritos and Cheetos, uses sunflower, corn, and canola oils, which contain mono- and polyunsaturated fats.

Most packaged foods have a label like this. It show how much and what kind of fats the food contains. These cornflakes contain no trans fats. The label would have to say if they did.

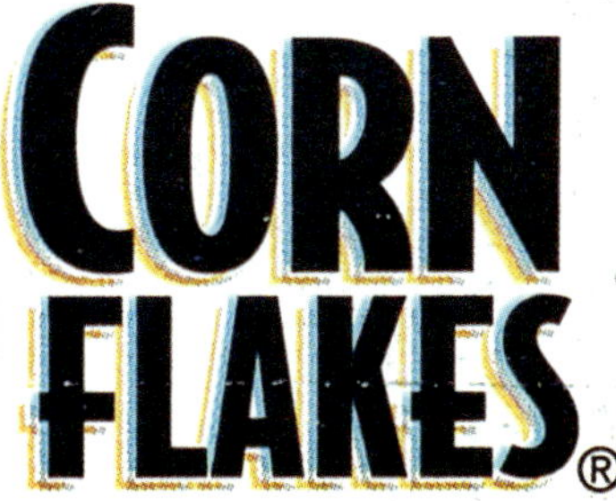

- **FAT FREE**
- **CHOLESTEROL FREE FOOD**

Nutrition Facts

Serving Size 1 Cup (28g/1.0 oz.)
Servings per Package About 18

Amount Per Serving	Cereal	Cereal with 1/2 Cup Vitamins A&D Fat Free Milk
Calories	100	140
Calories from Fat	0	0
	% Daily Value**	
Total Fat 0g*	0%	0%
Saturated Fat 0g	0%	0%
Cholesterol 0mg	0%	0%
Sodium 200mg	8%	11%
Potassium 25mg	1%	7%
Total Carbohydrate 24g	8%	10%
Dietary Fiber 1g	4%	4%
Sugars 2g		
Other Carbohydrate 21g		
Protein 2g		
Vitamin A	10%	15%
Vitamin C	10%	10%

Hydrogenated vegetable oil (HVO) is a type of renewable diesel fuel created by adding hydrogen to animal or plant fat.

But examples of hydrogenation aren't just found in vending machines—the process is also used to create high-grade gasoline. This occurs when large molecules are broken down into smaller molecules and react with hydrogen.

CALLING ALL CARBS!

In addition to hydrogenation, hydrogen can also bond with carbon and oxygen in a different process to create other fuels for our bodies called carbohydrates. These are any of the starches or sugars found in foods, especially in vegetables like corn and potatoes. Because pasta and bread come from plants like wheat, these foods also contain carbohydrates. Carbohydrates always contain carbon, hydrogen, and oxygen.

The process of digestion transforms the carbohydrates in your food into glucose. Glucose provides your cells with the energy they need to function.

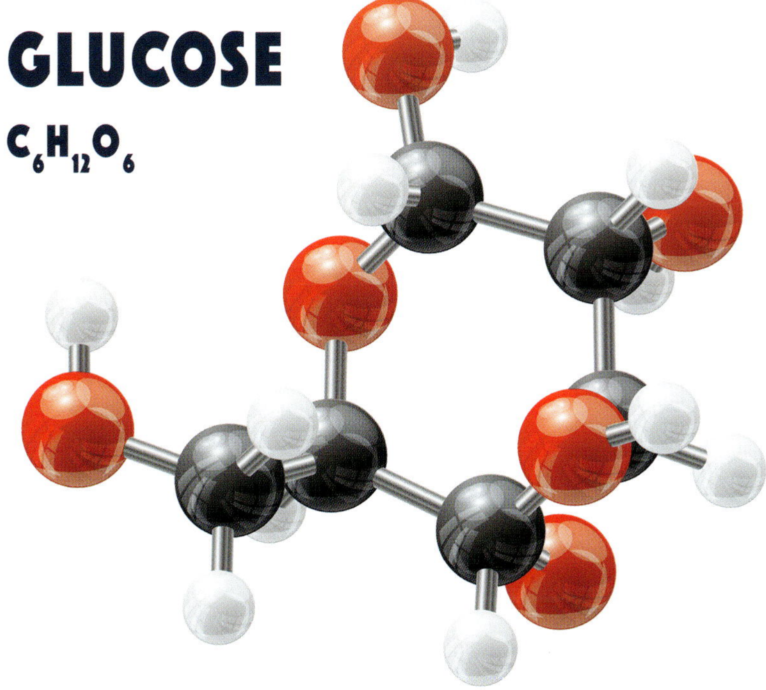

Foods rich in carbohydrates, such as potatoes and pasta, provide a lot of energy to the body. Many athletes will stock up on carbohydrates the night before a big competition.

Carbohydrates are created during a natural process called photosynthesis. This is when a plant absorbs carbon dioxide (CO_2) from the air, combines it with water it receives from its roots, and transforms it into either sugar ($C_6H_{12}O_6$) or starch ($C_6H_{10}O_5$).

Photosynthesis is the process by which plants use the light of the sun to transform water and carbon dioxide into carbohydrates—which is also fuel for whatever eats the plants. The word comes from Greek words that mean "to put together with the help of light."

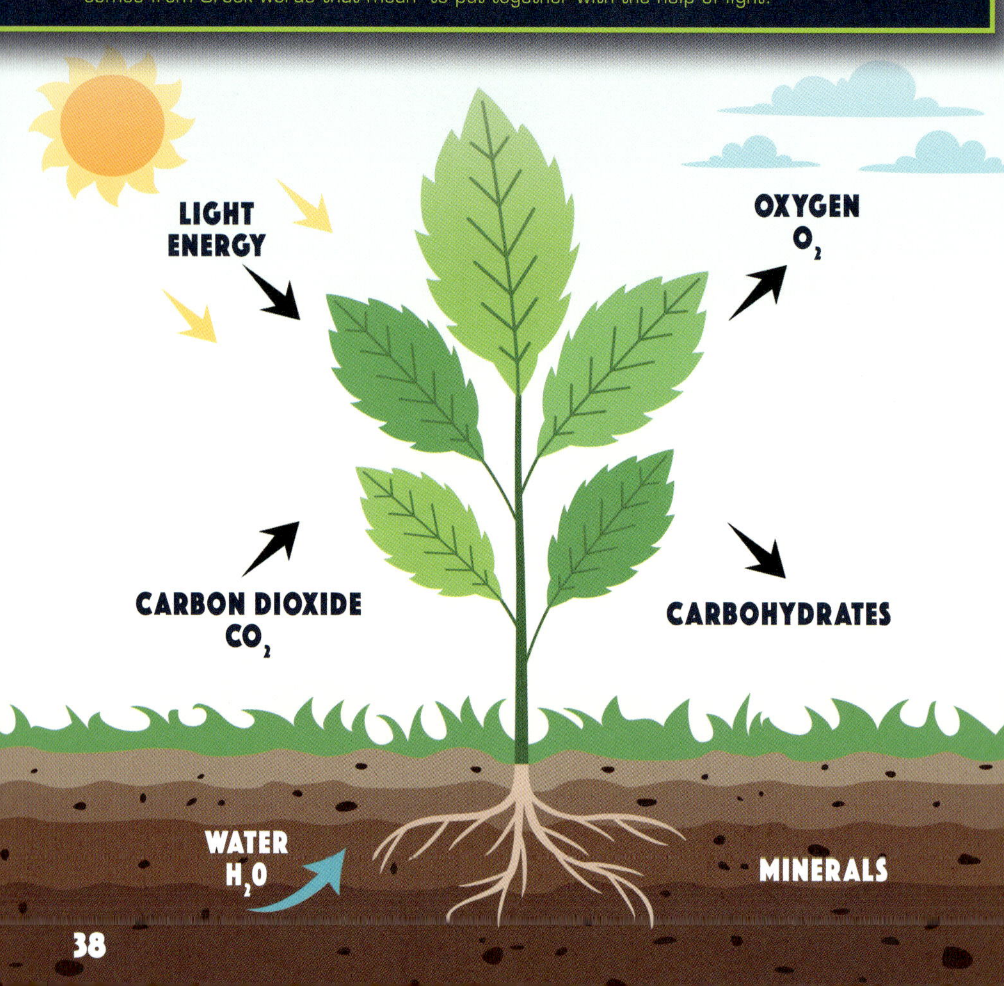

WHEN HYDROGEN MET CARBON

A hydrocarbon is an organic chemical compound that contains only hydrogen and carbon—no other elements. Many hydrocarbons occur naturally as part of plants. The chemicals that make leaves green and carrots orange are hydrocarbons. They're also found in fossil fuels like gas and oil. Crude rubber, plastic, explosives, and many industrial chemicals contain hydrocarbons.

It takes millions of years for coal to form deep in the earth. Mining coal is a dirty, dangerous business. The dust damages people's lungs, making them very sick. Some forms of mining also destroy the surrounding landscape. Human activity is depleting fossil fuels faster than the earth can replenish them. These are all good reasons for researching clean, efficient hydrogen fuel.

Hydrogen peroxide forms oxygen bubbles when it meets an enzyme called catalase. Catalase isn't found on the surface of your skin. It's in blood and bacteria. That's why bubbles form when peroxide is applied to a cut, but not when it comes in contact with unbroken skin.

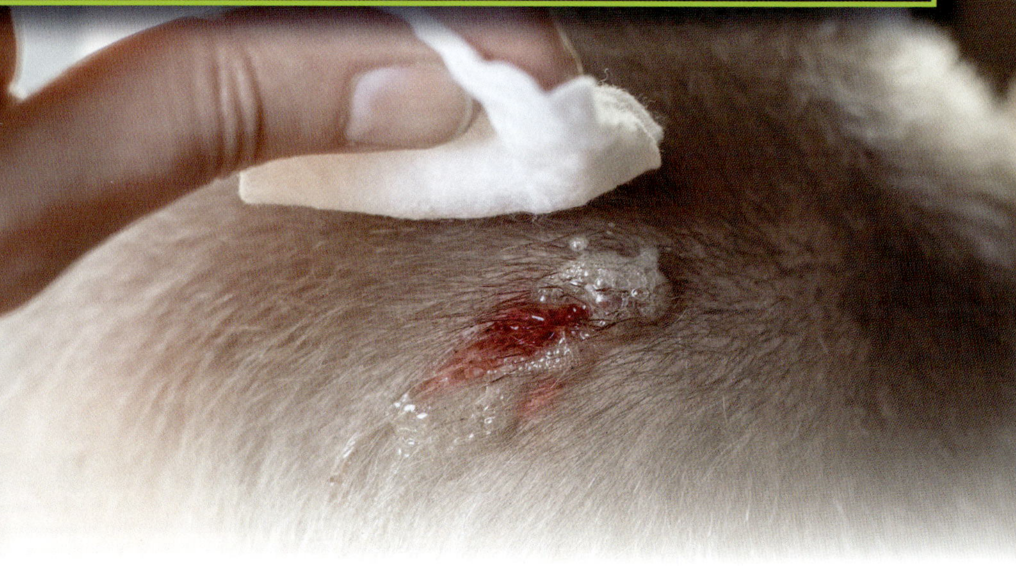

Nineteenth-century chemists divided hydrocarbons into two groups. Aliphatic hydrocarbons were derived from fats or oils. Aromatic hydrocarbons were derived from plants. Today, hydrocarbons are still called these two names, but their designation is based on their molecular structure.

Hydrocarbons are useful as fuel because hydrogen molecules release energy when exposed to heat. This is why coal, gas, and oil are used for heating homes. It's also why cars run on gasoline-powered engines. However, these fossil fuel sources have limits—people are using them up much faster than nature can replenish them. They also add to pollution. This is why many companies are devoting resources to learning how to use pure hydrogen for fuel.

Hydrogen compounds are all around us. They're found in the fuel that powers rockets to outer space and the ocean water that space capsules splash down in. The sugar ($C_{12}H_{22}O_{11}$) for your cereal is a carbon compound and so is the peroxide (H_2O_2) used to clean a scraped knee. There's even hydrogen compounds in your stomach that help you digest your food.

NASA is the largest consumer of liquid hydrogen in America.

CHAPTER 5

HYDROGEN POWERED

We know that hydrogen is very flammable, and pure hydrogen reacts with air, releasing energy as heat. We also know that the sun is a huge, burning ball of mostly hydrogen. So how does the heat from the sun reach all the way to the earth?

Scientists puzzled over this question for a long time. In the early 20th century, a German scientist named Albert Einstein (1879–1955) proposed theories to answer this question. Einstein theorized that matter could be converted to energy, and energy could be converted to matter. This is represented by his famous equation $E=MC^2$. He did not suggest how this would be accomplished, only that it was possible, given the right circumstances.

Those circumstances are met in the core of stars, like our sun, where a process called nuclear fusion occurs. Converting matter to energy releases a tremendous amount of power. Scientists estimate that converting the mass of a single paperclip could produce the same amount of energy as burning 15,000 barrels of oil. While that much energy is useful when you're talking about broadcasting heat and light across the galaxy, it's not practical for use on Earth. Releasing that much energy at once would result in a devastating explosion. However, under safely controlled circumstances, nuclear power could be a clean energy solution to replace fossil fuels.

THE THEORY OF RELATIVITY

Let's look at Einstein's famous equation a little more closely:

E	ENERGY
m	MASS (A PROPERTY OF MATTER)
c	SPEED OF LIGHT (186,282 MILES PER SECOND, OR 299,792 KM PER SECOND); THE SPEED AT WHICH LIGHT TRAVELS IN A VACUUM, WHICH IS AN AREA WITHOUT ANY MATTER
2	SQUARED (A NUMBER MULTIPLIED BY ITSELF)

What Einstein's theory proposes is that the laws of physics don't change. However, the same events would look very different to people moving at different speeds.

Brian May, co-founder and guitarist of the rock band Queen, is also a trained astrophysicist. He wrote a song called *'39* about astronauts experiencing the phenomena of time dilation. That is, they traveled near the speed of light for a year, but returned home to find a century had passed.

FUSION

So, what does $E=MC^2$ mean? To answer this, we must first learn about fusion. When four hydrogen atoms bang into each other, they eventually will fuse together to form a single helium atom. These collisions and the joining together of elements that result are called fusion. This process of fusion to form helium releases a lot of energy, which causes the sun to burn so hot. The theory of relativity explains how energy is released from fusion—enormous amounts of mass under a lot of pressure and extremely high temperatures create energy. According to the American Museum of Natural History (AMNH), fusion occurs on the sun "100 million quadrillion quadrillion times each second, and the sun has enough hydrogen to continue burning for another 5 billion years."

Nuclear fission is the process of splitting atoms to release energy. Nuclear fusion is the process of joining two light atoms to form one heavier atom while releasing energy. The process of nuclear fission is more dangerous, as it produces harmful radiation.

NUCLEAR FISSION vs. NUCLEAR FUSION

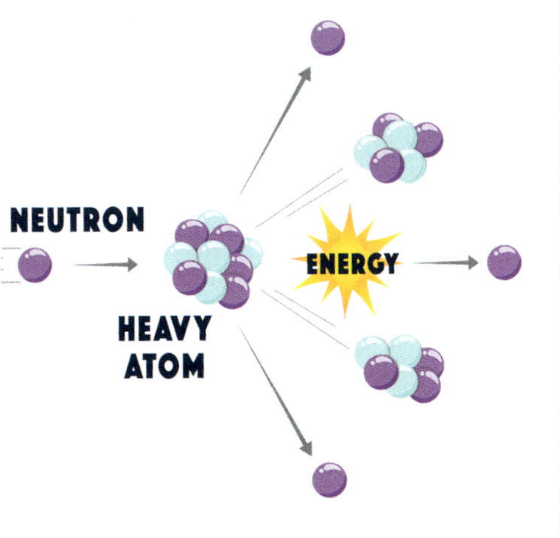

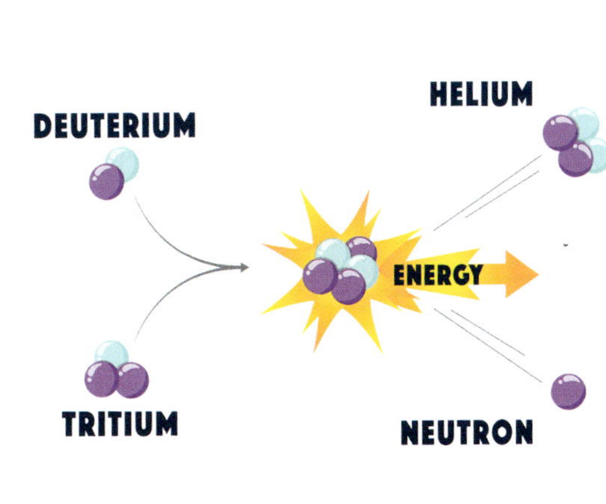

THE EXPLOSIVE POWER OF HYDROGEN

In the 1950s, American scientists attempted to harness the sun's energy when they invented the hydrogen bomb. The United States had already made the atom bomb, which was dropped on the cities of Hiroshima and Nagasaki, Japan, during World War II. These atom bombs were devastating, killing hundreds of thousands of people and ending the war.

Dubbed the H-bomb, the hydrogen bomb would be even more powerful. The atom bomb and other similar bombs used fission—the process of splitting atoms. The hydrogen bomb uses fusion, much like the sun does, to create enormous amounts of energy. Just how much energy this bomb contained would be seen in 1952, when it was dropped on a small island in the Pacific Ocean during an experiment. This bomb was 700 times more powerful than the atomic bomb that was dropped on Hiroshima. During the experiment, the H-bomb completely vaporized this island that had previously measured 1 mile (1.6 km) in diameter. It would take more than a million tons of dynamite to make a blast that big. According to reports of those who witnessed the blast, the bomb created a fireball that was 3 miles (5 km) wide, and a gigantic mushroom cloud rose from the blast in just 90 seconds. The power of the sun had been harnessed that day and would change world politics forever. Today, several countries, including the United States, possess hydrogen bombs.

POWERING THE FUTURE

Aside from bombs, though, scientists are also developing ways to harness the power of hydrogen to create a cleaner form of energy. Gasoline and other fossil fuels that we use as energy sources pollute the air. Several car manufacturers have already built hydrogen-powered cars and are rolling them out to customers. Along with hydrogen being a cleaner energy source, it is also recyclable. In other words, you can use parts of it again and again, unlike fossil fuels such as natural gas or coal, which are available only in a limited amount.

Bikini Atoll in the Marshall Islands was the site of several nuclear tests in the 1940s and 1950s. A hydrogen bomb tested in 1954 produced a much larger blast than expected, destroying three islands and producing a massive crater in the bottom of the lagoon. Due to lingering radiation from these tests, the area is still considered unsafe for people.

People didn't invent the concept of "Reduce, Reuse, Recycle." Nature has been doing it since the beginning of time!

THE HYDROGEN CYCLE

The hydrogen cycle shows how hydrogen can be cleanly used and reused. The hydrogen cycle starts with water, which is then split apart into hydrogen and oxygen using electrolysis. The oxygen is then released into the air, but the hydrogen is captured and contained in a fuel cell, which stores energy. These fuel cells can power many things, such as school buses, airplanes, and space shuttles. However, unlike cars and planes that run on fossil fuels, when the hydrogen fuel cells are used in these machines, they create very little (and sometimes no) pollution.

The main thing that is released from these fuel cells is water, which can then be split into hydrogen and oxygen, and the hydrogen can be used all over again. Being able to recycle hydrogen may change the way we live.

CRYOGENICS

Because of the hydrogen bomb, the sun, the *Hindenburg*, and all the fuels derived from hydrogen, it is easy to think only of fire when we think of this element. But hydrogen can also be found in a very cold form—liquid hydrogen. Liquid hydrogen is so incredibly cold that if you were to place a tennis ball in a bowl of liquid hydrogen, take it out, and then immediately bounce it on the ground, the ball would shatter like glass.

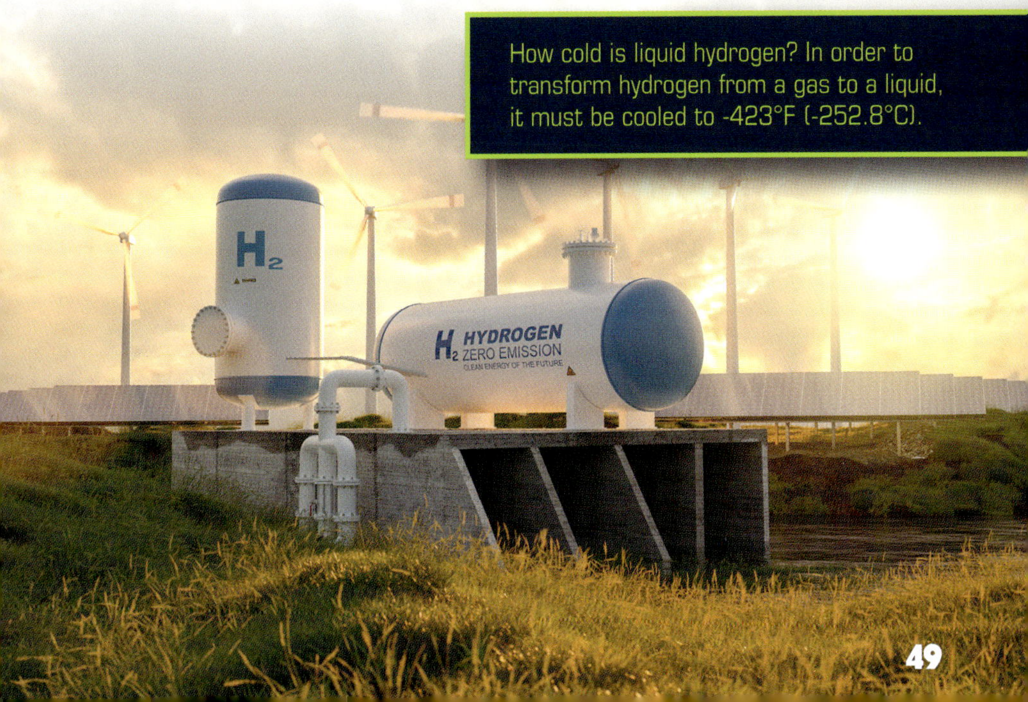

How cold is liquid hydrogen? In order to transform hydrogen from a gas to a liquid, it must be cooled to -423°F (-252.8°C).

Liquid hydrogen is cold enough to cause severe frostbite, yet it's also extremely flammable.

Liquid hydrogen is used in cryogenics, which is the study of the effects of very cold temperatures. The National Aeronautics and Space Administration (NASA) is one agency that is studying cryogenics. For example, when a space shuttle, satellite, or another spacecraft is returning to Earth from space, it is moving at a tremendous speed. Entering Earth's atmosphere at tremendous speeds can create extremely high temperatures, which could severely damage the spacecraft. Cryogenics can help engineers and other researchers learn how to properly cool down such vessels during reentry.

At the NASA Goddard Space Flight Center in Greenbelt, Maryland, scientists are using cryogenics to observe stars. According to NASA's website, "Sensitive sensors can catch even the weakest signals reaching us from the stars. Many of these sensors must be cooled well below room temperature to have the necessary sensitivity."

NASA has used liquid hydrogen to fuel spacecraft since the beginning of the U.S. space program. Kennedy Space Center in Florida is home to the world's largest liquid hydrogen fuel tanks and transfer system.

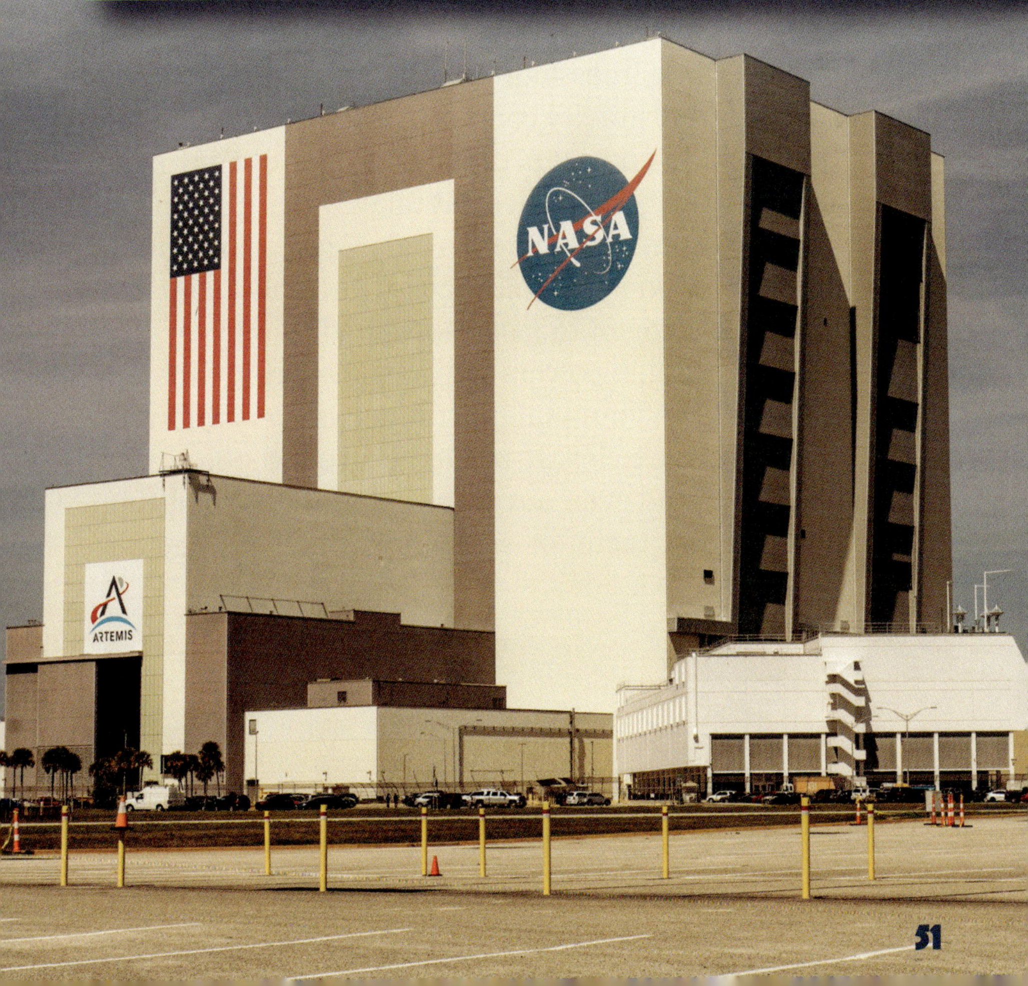

WHAT MAKES THE SUN SHINE?

Like all the other stars we know, the sun is mostly composed of hydrogen and helium. Hydrogen accounts for about 90 percent of the sun's atoms and 70 percent of the sun's mass. Helium accounts for a little less than 10 percent of the atoms and 30 percent of the sun's mass. The rest is composed of trace amounts of other elements such as oxygen, carbon, nitrogen, silicon, magnesium, neon, iron, and sulfur.

In the sun's core, intense heat—about 27 million degrees Fahrenheit (15 million degrees Celsius)—and gravity compress hydrogen atoms into helium. The energy produced travels out into the universe as light and heat. It takes about 8 minutes for light from the sun to reach Earth. The sun is about 4.5 billion years old. Scientists think it contains enough hydrogen to last another 5 billion years.

Hydrogen powers the universe. From the stars to the food on your plate, hydrogen is everywhere. It powers the vehicles we use to get around, the plants that grow, and even our own bodies. For something we can't see, taste, or smell, hydrogen sure gets a lot done!

Hydrogen powers the sun, giving our planet the light and heat necessary for life.

THE PERIODIC TABLE OF ELEMENTS

Hydrogen powers the universe. From the earliest times, people have harnessed hydrogen to power industry—from mills driven by waterwheels to rocket engines. Hydrogen does it all.

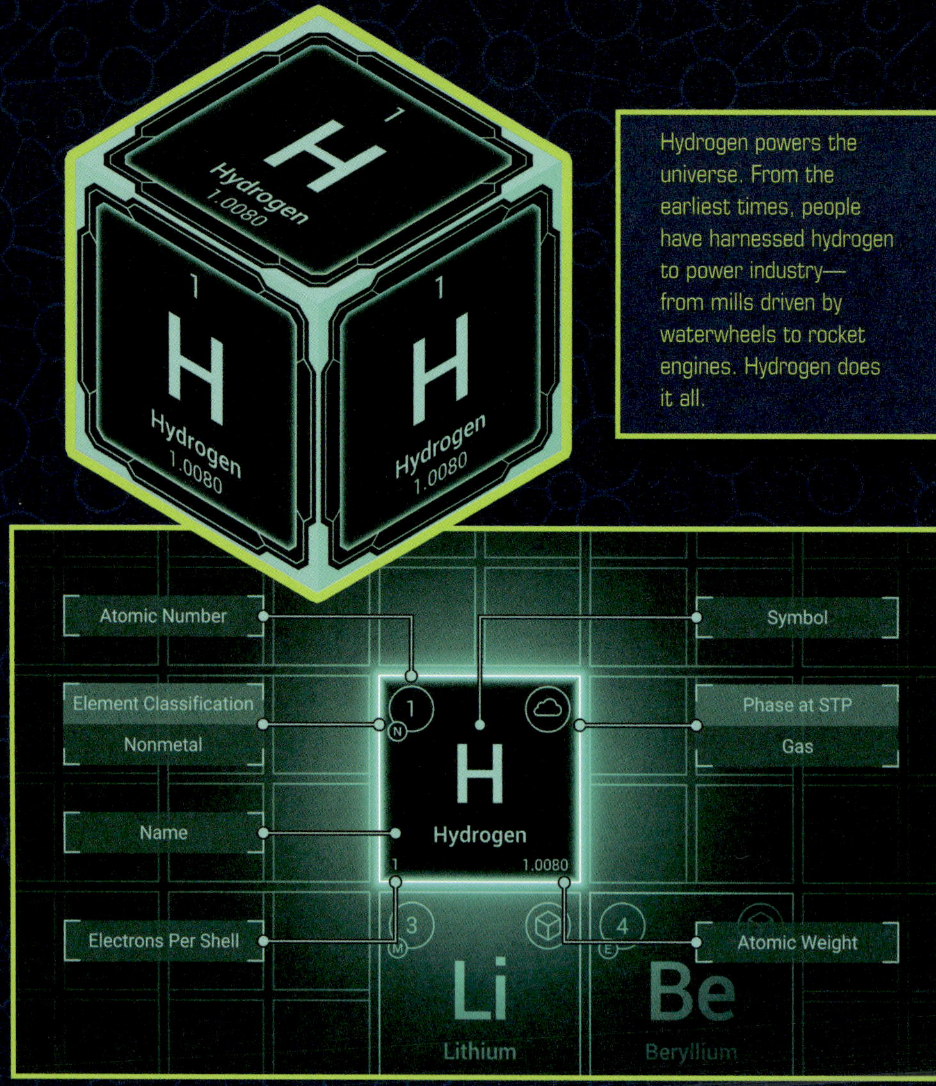

PERIODIC TABLE OF ELEMENTS

- Alkali metals
- Alkaline earth metals
- Transition metals
- Metal
- Metalloids
- Nonmetals
- Halogen
- Noble gases
- Lanthanoids
- Actinoids

1	2	3	4	5	6	7	8	9	10	11	12	13	14	15	16	17	18
1 H Hydrogen 1.008																	2 He Helium 4.0026
3 Li Lithium 6.94	4 Be Beryllium 9.0122											5 B Boron 10.81	6 C Carbon 12.011	7 N Nitrogen 14.007	8 O Oxygen 15.999	9 F Fluorine 18.998	10 Ne Neon 20.180
11 Na Sodium 22.990	12 Mg Magnesium 24.305											13 Al Aluminium 26.982	14 Si Silicon 28.085	15 P Phosphorus 30.974	16 S Sulfur 32.06	17 Cl Chlorine 35.45	18 Ar Argon 39.948
19 K Potassium 39.098	20 Ca Calcium 40.078	21 Sc Scandium 44.956	22 Ti Titanium 47.867	23 V Vanadium 50.942	24 Cr Chromium 51.996	25 Mn Manganese 54.938	26 Fe Iron 55.845	27 Co Cobalt 58.933	28 Ni Nickel 58.693	29 Cu Copper 63.546	30 Zn Zinc 65.38	31 Ga Gallium 69.723	32 Ge Germanium 72.630	33 As Arsenic 74.922	34 Se Selenium 78.971	35 Br Bromine 79.904	36 Kr Krypton 83.798
37 Rb Rubidium 85.468	38 Sr Strontium 87.62	39 Y Yttrium 88.906	40 Zr Zirconium 91.224	41 Nb Niobium 92.906	42 Mo Molybdenum 95.95	43 Tc Technetium (98)	44 Ru Ruthenium 101.07	45 Rh Rhodium 102.91	46 Pd Palladium 106.42	47 Ag Silver 107.87	48 Cd Cadmium 112.41	49 In Indium 114.82	50 Sn Tin 118.71	51 Sb Antimony 121.76	52 Te Tellurium 127.60	53 I Iodine 126.90	54 Xe Xenon 131.29
55 Cs Caesium 132.91	56 Ba Barium 137.33	57-71	72 Hf Hafnium 178.49	73 Ta Tantalum 180.95	74 W Tungsten 183.84	75 Re Rhenium 186.21	76 Os Osmium 190.23	77 Ir Iridium 192.22	78 Pt Platinum 195.08	79 Au Gold 196.97	80 Hg Mercury 200.59	81 Tl Thallium 204.38	82 Pb Lead 207.2	83 Bi Bismuth 208.98	84 Po Polonium (209)	85 At Astatine (210)	86 Rn Radon (222)
87 Fr Francium (223)	88 Ra Radium (226)	89-103	104 Rf Rutherfordium (267)	105 Db Dubnium (268)	106 Sg Seaborgium (269)	107 Bh Bohrium (270)	108 Hs Hassium (277)	109 Mt Meitnerium (278)	110 Ds Darmstadtium (281)	111 Rg Roentgenium (282)	112 Cn Copernicium (285)	113 Nh Nihonium (286)	114 Fl Flerovium (289)	115 Mc Moscovium (290)	116 Lv Livermorium (293)	117 Ts Tennessine (294)	118 Og Oganesson (294)

57 La Lanthanum 138.91	58 Ce Cerium 140.12	59 Pr Praseodymium 140.91	60 Nd Neodymium 144.24	61 Pm Promethium (145)	62 Sm Samarium 150.36	63 Eu Europium 151.96	64 Gd Gadolinium 157.25	65 Tb Terbium 158.93	66 Dy Dysprosium 162.50	67 Ho Holmium 164.93	68 Er Erbium 167.26	69 Tm Thulium 168.93	70 Yb Ytterbium 173.05	71 Lu Lutetium 174.97
89 Ac Actinium (227)	90 Th Thorium 232.04	91 Pa Protactinium 231.04	92 U Uranium 238.03	93 Np Neptunium (237)	94 Pu Plutonium (244)	95 Am Americium (243)	96 Cm Curium (247)	97 Bk Berkelium (247)	98 Cf Californium (251)	99 Es Einsteinium (252)	100 Fm Fermium (257)	101 Md Mendelevium (258)	102 No Nobelium (259)	103 Lr Lawrencium (266)

- C Solid
- Hg Liquid
- H Gas
- Rf Unknown

GLOSSARY

atom: The smallest, most basic unit of an element; made up of protons, neutrons, and electrons.
atomic number: The number of protons in the nucleus of an atom, which is usually equal to the number of electrons.
atomic weight: The mass of one atom of an element.
cryogenics: The study of extremely low temperatures.
diatomic: Consisting of two atoms of the same element.
electron: A negatively charged particle outside an atom's nucleus.
element: The basic matter that all things are made of; any matter made up of only one kind of atom.
fission: The splitting of an atomic nucleus.
flammable: Capable of catching fire easily.
fusion: The coming together of atomic nuclei to form heavier nuclei, resulting in the production of large amounts of energy.
hydrogenation: The process of adding hydrogen to a molecule; often used in producing food.
molecule: The smallest bit of matter before it gets broken down into its basic parts, or atoms.
neutron: A particle within the nucleus of an atom that contains no charge; found in the nucleus of all elements except hydrogen.
nomenclature: A system or set of terms or symbols especially in a particular science, discipline, or art.
nucleus: The positively charged central portion of an atom.

period: A horizontal row in the periodic table.

photosynthesis: The formation of carbohydrates from carbon dioxide and hydrogen (typically from the sun).

pollution: The action of making an environment unsuitable or unsafe for use by introducing man-made waste.

proton: A positively charged particle within the nucleus of an atom. The number of protons and electrons is almost always equal.

radioactivity: The property possessed by some elements (such as uranium) or isotopes (such as carbon 14) of spontaneously emitting energetic particles (such as electrons or alpha particles) by the disintegration of their atomic nuclei.

recycle: In nature, to reuse or make (a substance) available for reuse for biological activities through natural processes of biochemical degradation or modification.

theory: A plausible or scientifically acceptable general principle or body of principles offered to explain phenomena.

FOR MORE INFORMATION

AMERICAN MUSEUM OF NATURAL HISTORY

200 Central Park West
New York, NY 10024-5102
(212) 769-5000
Website: www.amnh.org
Facebook: @naturalhistory
Instagram and X: @amnh
AMNH is a world-class museum dedicated to advancing research and sharing scientific knowledge about the natural world.

CHEMICAL INSTITUTE OF CANADA (CIC)

90-2420 Bank Street
Ottawa, Ontario, Canada
K1V 8S1
(613) 232-6252
Website: www.cheminst.ca/
X: @CIC_ChemInst
The CIC facilitates the research and cooperation of chemists and chemical engineers in Canada.

FUEL CELL & HYDROGEN ENERGY ASSOCIATION (FCHEA)

1025 Connecticut Avenue Northwest, Suite 1000
Washington D.C. 20036
(202) 355-9463
Website: www.fchea.org/
Facebook: @FCHEA
X: @FCHEA_News
FCHEA is an American organization dedicated to advancing commercialization and finding markets for fuel cells and hydrogen energy.

INTERNATIONAL UNION OF PURE AND APPLIED CHEMISTRY (IUPAC)

IUPAC Secretariat
79 T.W. Alexander Drive
Research Commons Building 4501, Suite 190
Research Triangle Park, NC 27709, USA
(919) 485 8700
Website: iupac.org/
Facebook: @iupac.org
X: @iupac
IUPAC is the global chemistry organization that maintains the periodic table of elements.

INTERNATIONAL UNION OF PURE AND APPLIED PHYSICS (IUPAP)

Fondazione Internazionale Trieste per il Progresso e la Libertà delle Scienze
c/o ICTP
Strada Costiera 11
34151 Trieste
Website: iupap.org/
Facebook: @iupap
X: @iupap_physics
IUPAP's mission is to assist in the global advancement of physics and facilitate international scientific cooperation.

LOS ALAMOS NATIONAL LABORATORY

PO Box 1663
Los Alamos, NM 87545
(505) 667-5061
Website: www.lanl.gov/
Facebook: @LosAlamosNationalLab
Instagram and X: @losalamosnatlab
Los Alamos National Laboratory is a federally funded research and development center dedicated to solving national security issues with science.

FOR FURTHER READING

Congdon, Lisa. *The Illustrated Encyclopedia of the Elements: The Powers, Uses, and Histories of Every Atom in the Universe.* San Fransico, CA: Chronicle Books, LLC, 2021.

DeVault, Lee. T*he Adventures of Hydrogen and Oxygen: The Creation of Water.* Independently Published, 2023.

DK. *Super Simple Chemistry: The Ultimate Bitesize Study Guide.* New York, NY: DK Publishing, 2020.

McHenry, Ellen Johnston. *The Chemical Elements Coloring & Activity Book: A Fun and Interactive Guide to the World of Chemistry.* Independently Published, 2021.

Silver, Donald, and Patricia Wynne. *My First Book About Chemistry.* Mineola, NY: Dover Publications, 2020.

Thomas, Isabel, and Sara Gillingham. E*xploring the Elements: A Complete Guide to the Periodic Table.* New York, NY: Phaidon Press, Inc., 2021.

Woodbury, Rebecca. *Atoms and Molecules Meet.* Real Science-4-Kids, 2024.

Zovinka, Edward P., and Rose A. Clark. *A Kids' Guide to the Periodic Table: Everything You Need to Know About the Elements.* Emeryville, CA: Rockridge Press, 2020.

INDEX

ABOUT THE AUTHOR

KATHLEEN A. KLATTE is the author of many nonfiction books for children and teens. Topics she has written about range from animals and nature to constitutional law to unusual career choices. She lives in New York with one cat and far too many books and Legos.

PHOTO CREDITS

Cover, p. 1 remotevfx.com/Shutterstock.com; Cover, p. 1 pluie_r/ Shutterstock.com; pp. 3-64 Oksancia/Shutterstock.com; pp. 5, 47 Everett Collection/Shutterstock.com; p. 7 Scharfsinn/Shutterstock.com; p. 9 German Vizulis/Shutterstock.com; p. 10 Corona Borealis Studio/ Shutterstock.com; p. 12 Double Brain/Shutterstock.com; p. 13 AP vector/ Shutterstock.com; p. 15 Ipajoel/Shutterstock.com; p. 17 Srg Gushchin/ Shutterstock.com; p. 18 Stylooze_Art/Shutterstock.com; p. 19 Olga Popova/ Shutterstock.com; p. 20 Eugene B-sov/Shutterstock.com; p. 23 Zakharchuk/Shutterstock.com; p. 25 Sheila Fitzgerald/ Shutterstock.com; p. 26 Proonty/Shutterstock.com; p. 27 Akarawut/ Shutterstock.com; p. 29 https://commons.wikimedia.org/wiki/ File:HD.3F.003_(11086395496).jpg; p. 31 ZikG/Shutterstock.com; p. 32 Krakenimages.com/Shutterstock.com; p. 33 luchschenF/ Shutterstock.com; p. 35 Dennis MacDonald/Shutterstock.com; p. 36 3rdtimeluckystudio/Shutterstock.com; p. 37 dip/Shutterstock.com; p. 28 yusufdemirci/Shutterstock.com; p. 39 Maksim Safaniuk/ Shutterstock.com; p. 40 Natalia Kokhanova/Shutterstock.com; p. 41 Paitoon Pornsuksomboon/Shutterstock.com; p. 44 https://commons. wikimedia.org/wiki/File:Brian_May_(NHQ201812310024).jpg; p. 45 Julee Ashmead/Shutterstock.com; p. 48 dee karen/ Shutterstock.com; p. 49 r.classen/Shutterstock.com; p. 50 sruilk/ Shutterstock.com; p. 51 Nadezda Murmakova/Shutterstock.com; p. 53 titoOnz/Shutterstock.com; p. 54 ShannonChocolate/ Shutterstock.com; p. 55 View_From_My_Lens/Shutterstock.com.